Turtle Story

by Sharon Kahkonen

Orlando Austin New York San Diego Toronto London

Visit *The Learning Site!*
www.harcourtschool.com

Introduction

A warm breeze drifts over the ocean. Waves gently wash against the beach. In the darkness, a female loggerhead sea turtle swims toward the shore. Her large flippers and streamlined body help her swim with ease. Her front flippers are like the wings of an airplane. They provide lift and the force she needs to move through the water. She moves up and down in the water with little effort. She can swim fast.

She is amazing to look at! Her heart-shaped shell is 92 centimeters (37 in.) long. It is reddish-brown in color. The shell protecting her underside is creamy yellow. She has a large, blocklike head and powerful jaws. Her neck is short and broad. Her body weighs 115 kilograms (254 lb).

Loggerhead turtles are graceful swimmers.

Nesting

Now she reaches the beach. Although she can swim quickly in the ocean, she is slow on land. She uses her flippers to pull herself. She has to drag her body over the sand. It is tiring work. Why would she want to be on land in the first place? She has left the sea to build a nest and lay eggs.

She pauses often as she makes her way up the beach. She is not well adapted for the land. She cannot pull her head and legs into her shell for protection. It is a quiet night, which is good because light and human activity might cause her to stop nesting and leave. She stops at a dry spot above the high water mark. It is only a few hundred meters from where she hatched about 30 years ago. In those 30 years, she has traveled hundreds of miles at sea. But she comes back to this same Florida beach to lay her eggs.

Female loggerhead turtles build their nests on land.

She needs to get to the high water mark to build her nest. If it were lower, the nest would flood when the tide came in.

She starts flinging away loose sand with her front flippers to dig a pit. When she finishes, she digs out an egg cavity with her back flippers. The egg cavity is about 20 centimeters (8 in.) wide and 46 centimeters (18 in.) deep.

When the egg cavity is complete, she lays her eggs. They are round, white, and about the size of a golf ball. The eggs are leathery and flexible. They do not break as they fall into the egg cavity.

She lays almost 100 eggs! Then, she uses her rear flippers to push sand over the top of the nest. She throws sand in all directions to disguise the nest. A hundred eggs would make a nice meal for many predators! Her work is now done. She makes her way back to the sea. She never comes back to check on her nest or her eggs. The eggs, and the young turtles inside them, are on their own.

The Eggs Develop and Hatch

Luckily, humans do not disturb this stretch of the Florida coast too much. It has been set aside as a safe place for nesting sea turtles.

For 55 days, the young loggerheads grow and develop inside their eggs. Finally, they are ready to hatch. Each turtle begins to tap at the inside of its shell using its sharp egg tooth. It is hard work, but the young turtles finally break through their shells. The turtles are only 5 centimeters (2 in.) long! They do not have hard, protective shells yet.

After hatching, the young turtles work together to dig out of their nest. This takes several days. Then, one night, they all burst out of the sand. They race as quickly as they can to the ocean. Hungry gulls and crabs are waiting for this moment! Predators eat many young turtles during their short trip from the nest to the ocean.

In fact, scientists think that only about one in 1,000 to one in 10,000 hatchlings make it to adulthood.

Let's follow one tiny female hatchling that does grow up to become an adult.

Newly hatched loggerhead turtles racing toward the ocean

Swimming Out to the Open Ocean

Our female hatchling makes it safely to the ocean. She swims into the waves, which lead her into the open ocean. It may not seem that the open ocean would be safe. But compared with the shore, it is safer.

She swims for several miles. Finally, she reaches the Gulf Stream. The Gulf Stream is a huge current of warm water. It moves north along the eastern coast of North America. Here, she begins the next part of her life. She will swim around in ocean currents for years. There, she will eat and grow larger.

The turtles that make it to the sea may be lucky enough to survive and spend the next years of their lives in the currents of the North Atlantic gyre.

The Life of Young Loggerheads

The Gulf Stream is part of a circular current system. This system is called the North Atlantic gyre. It circles clockwise around the northern Atlantic Ocean. The North Atlantic gyre surrounds the Sargasso Sea. The North Atlantic gyre has a lot of algae, called *Sargassum,* growing in it and floating on the surface. There are many organisms living among the *Sargassum.* These organisms include small sea animals, such as tiny sponges and jellyfishes.

The young loggerhead stays within the North Atlantic gyre for 10 years. She floats and swims among the *Sargassum.* She feeds on the *Sargassum* and other organisms living in the algae.

She has to be sure to stay within the gyre, where the water is warm and the food is plentiful. If she drifts too far north, she may get caught in very cold water. She could easily die from the cold. If she strays too far south, she could be carried far from the loggerhead's normal range.

The North Atlantic gyre is not without its dangers. Birds are attracted to the animal life in the seaweed. They could grab an unsuspecting turtle. Other floating objects, including garbage, get caught up in the North Atlantic gyre. There are plastic bags and balloons. Young turtles can mistake these objects for food. When a young turtle swallows plastic, it gets caught in the turtle's stomach. The turtle does not get enough nutrition and it starves to death.

Our turtle is a lucky one. She survives to the age of 10. She has grown to about the size of a dinner plate. Now it is time for her to change her way of life.

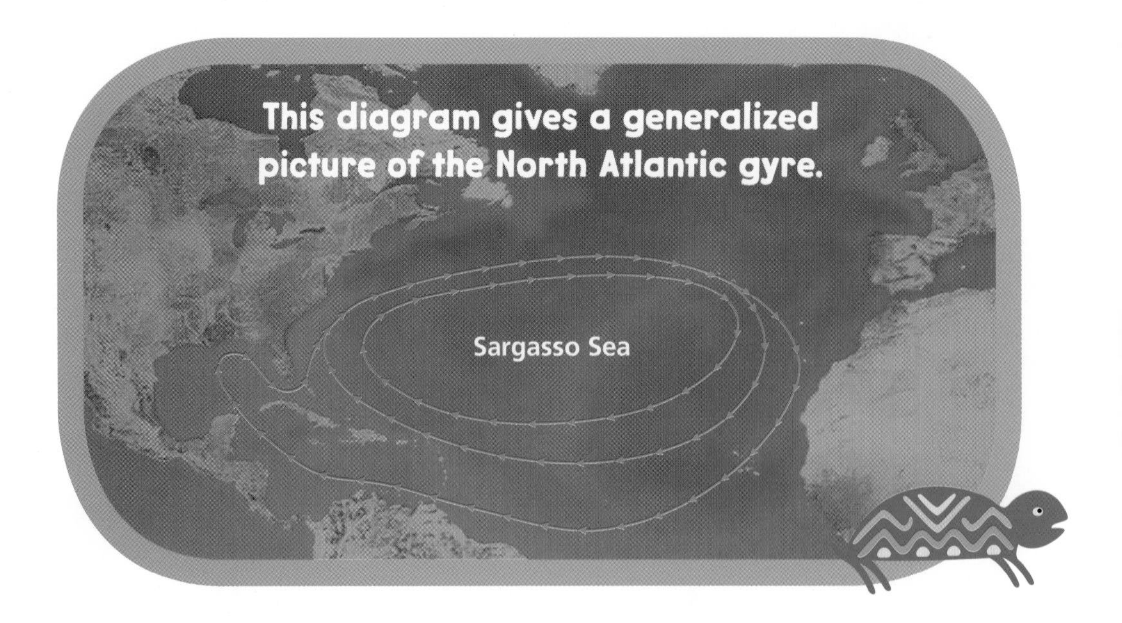

This diagram gives a generalized picture of the North Atlantic gyre.

She moves to shallower waters, and she changes her diet. She has migrated to the shallow, coastal areas in the western Atlantic. She feeds on the floor of lagoons, bays, river mouths, and shallow coastal waters.

On a typical day, she spends her time feeding and resting. She feeds on crabs, shrimp, jellyfish, and a variety of mollusks. Her beak-like jaws are strong enough to crush the shells of her prey.

For the next 10 years, she swims and feeds up and down the coast. She has now reached adulthood. At about 20 years old, she has grown to a length of 51 centimeters (20 in.). At this size, she is safe from many ocean predators. But, she still has to watch out for large fish, such as tiger sharks.

Adult loggerhead feeding and swimming

Return to Mating Grounds

Twenty years ago, she hatched on a Florida beach. She is now swimming toward the same beach. It is mating season. Along the way, she mates. Finally, she is ready to lay her eggs.

Like her mother before her, she swims to shore. She pulls herself along the beach. This is the first time she has touched land since hatching! When she finishes laying her eggs, she returns to the ocean. She will continue to return to this beach several more times during her life to lay eggs.

But for now, a new generation of loggerheads lies buried beneath the sand. Each one is getting ready to begin a new life. For the lucky few, it will be a long journey to adulthood.

Female loggerheads return to the ocean after laying their eggs.

What Is Happening to the Loggerheads?

Loggerhead turtles are becoming rare. This is because they are threatened. Being threatened means that there are few animals of this type left. The few that are left are decreasing in number.

There are many threats to the survival of loggerheads. Most of these threats are caused by humans. For example, people throughout history have killed loggerheads for their meat, eggs, shells, and skin. We need to protect these turtles and their nesting grounds.

There is a good chance that loggerhead nesting beaches could be destroyed. Many female loggerheads use Florida beaches for nesting. These same beaches attract humans. More and more people are moving to Florida every day. The beaches are being developed for homes and recreation. Artificial lights and increased traffic are among the many things that make these beaches difficult for turtle nesting.

Polluted waters can make loggerhead turtles sick. People dump sewage, oil, and other chemicals into the ocean. These chemicals cause diseases that can kill turtles.

Turtles might accidentally get caught in fishing gear or shrimp nets. Fishing boats use long lines and drift nets in places where big ocean currents meet. Algae, jellyfish, and other small sea animals collect here. This attracts larger sea animals like fish and sea turtles. The fish attract fishing boats. Sea turtles can get caught in the fishing gear. They might drown or get hurt.

What Are We Doing to Help?

There are now regulations to help stop sea turtles from getting caught in the nets used to catch shrimp. Boats that use nets to catch shrimp must use TEDs. TEDs are Turtle Exclusion Devices. They allow turtles to escape if they enter the nets. Also, some countries are passing laws to help protect the turtles. The United States is one of these countries.

Loggerhead sea turtles are beautiful creatures. They have lived in our oceans for hundreds of thousands of years. They are part of Earth, just like us. But to continue surviving, they will need all the help they can get.

Protecting turtle nests helps turtles survive at least until they hatch.